ABOUT TIME

Earth Time

Brian Williams

SMART APPLE MEDIA

First published by Cherrytree Press
(a member of the Evans Publishing Group)
327 High Street, Slough
Berkshire SL1 1TX, United Kingdom
Copyright © 2002 Evans Brothers Limited
This edition published under license from
Evans Brothers Limited.

Created and designed by
THE FOUNDRY DESIGN AND PRODUCTION
Crabtree Hall, Crabtree Lane, Fulham, London, SW6 6TY

Special thanks to Vicky Garrard, Colin Rudderham, and the late Helen Courtney

Published in the United States by
Smart Apple Media
1980 Lookout Drive
North Mankato, MN 56003

Printed in Hong Kong

Library of Congress Cataloging-in-Publication Data

Williams, Brian, 1943- Earth time / by Brian Williams.
p. cm. — (About time) Includes index. Summary: Explores the history
of time as it relates to the evolution of the Earth, from its origins as a lifeless
rock, through the development of simple organisms, more complex animals,
and, eventually, humans.
ISBN 1-58340-210-1
1. Historical geology—Juvenile literature. [1. Historical geology.
2. Evolution.] I. Title.

QE28.3 .W545 2002   551.7—dc21   2002023111

2 4 6 8 9 7 5 3 1

Acknowledgments
The author and publishers would like to thank the following for permission to reproduce
Photographs: Front cover and Title page Topham Picturepoint and Foundry Arts page 4 (top)
Topham Picturepoint (bottom) Simon Harris/Robert Harding Picture Library page 5 Tony
Waltham/Robert Harding Picture Library page 6 (left) Science Museum/Science & Society
Picture Library (right) Topham Picturepoint page 8 (top) Galaxy Pictures/Robin Scagell page 9
Tony Waltham/Robert Harding Picture Library page 10 Christie's Images/Bridgeman Art Library
page 11 (top) Science Museum/Science & Society Picture Library (bottom) Index
Stock/Phototake NYC/Robert Harding Picture Library page 12 (left) Topham Picturepoint (right)
Tony Waltham/Robert Harding Picture Library page 13 Topham Picturepoint page 14 (top) Tony
Waltham/Robert Harding Picture Library (bottom) Tom McHugh/Science Photo Library page 15
Martyn F.Chillmaid/Robert Harding Picture Library page 16 (top) David Scharff/Science Photo
Library (bottom) Topham Picturepoint page 17 John Reader/Science Photo Library page 18
(top) Roland Seitre/Still Pictures (bottom) Topham Picturepoint page 19 Tony Waltham/Robert
Harding Picture Library page 20 (all) Topham Picturepoint page 21 Chris Butler/Science Photo
Library page 22 (top) Topham Picturepoint (bottom) Tom McHugh /Science Photo Library page
23 Joe Tucciarone/Science Photo Library page 24 Topham Picturepoint page 25 Robert Harding
Picture Library page 26 (left) Topham Picturepoint (right) E.Simanor/Robert Harding Picture
Library page 27 Topham Picturepoint page 28 Musee d'Orsay, Paris, France/Bridgeman Art
Library page 29 (left) Gavin Hellier/Robert Harding Picture Library (right) Simon Harris/Robert
Harding Picture Library All graphics (8 and 24) are courtesy of Foundry Arts.

Contents

Introduction

HOW OLD IS THE EARTH?

Was there anything before Earth? How long ago did dinosaurs live? When did human beings first walk this planet? These are all questions about time, and the answers involve stretches of time so long that most of us find the numbers involved difficult to grasp.

We measure our daily activities in small portions of time—seconds, minutes, and hours. Our lives are measured in slightly longer chunks—days, weeks, months, and years. A lifetime for a human can be as long as 100 years, or sometimes slightly longer, but that is hardly any time at all. The pyramids of Egypt are 4,500 years old; they belong to history. Before history (when people wrote things down) is known as prehistory. When we think about prehistoric Earth, we must start thinking in millions, not thousands, of years.

Our planet is about 4.6 billion years old. Throughout its existence, the Earth's surface has been constantly changing. Time never stands still; nor does life.

HOW LONG IS EARTH TIME?

Our sun is a star, and the Earth is one of nine planets moving around this star. The Earth is an unusual planet because it has life, but how and when was it made? To find the answers, scientists study the remains of living things, the rocks of which the Earth is made, and the stars in the universe. They have worked out that our planet is about 4.6 billion years old; they believe the universe is much older—perhaps 13 billion years old.

COUNTING BACK TO CREATION

Before modern science began, people thought the world was much younger than it is. European scholars in the Middle Ages used dates in the Bible and history books to count back to Creation, the so-called beginning of the world. In 1654, Archbishop James Ussher worked out that God created the world on October 26, 4004 B.C.—at nine o'clock in the morning.

The Colorado River has gouged out the great gorge of the Grand Canyon in Arizona. The Canyon's oldest rocks are two billion years old.

In India, the Hindus had long believed in far longer cycles of time, each lasting more than four billion years. In the 18th century, scientists cast doubt on Archbishop Ussher's "Creation calendar." In France, Georges Buffon measured how long heated metal globes took to cool down, and concluded that a hot Earth would have taken 74,000 years to cool. Other scientists, such as the Scottish geologist James Hutton, began looking at what the rocks could tell us about time.

THE FOSSIL REVOLUTION

In the 19th century, people made exciting finds of fossil dinosaurs. These were the remains of animals no longer living. What had happened to them? As scientists began to discover a record of existence in the rocks, they found signs of life long before the dinosaurs. They realized that over millions of years, life had undergone many changes.

HOW THINGS BEGAN

Life on Earth began some time during the first 1.3 billion years of its existence. Tiny animals and plants developed into larger, more complex kinds, many of which died out. Others changed, taking the forms we know today. The process scientists call evolution took an immensely long time. Humans did not show up until around five million years ago, by which time an amazing variety of creatures had already crawled, swum, and flapped about the Earth—dying out without ever being seen by human eyes. We treat the planet as our own, but our time—human history—is no more than a brief sentence near the end of the ancient chronicles of Earth time.

The awesome bite of the Tyrannosaurus rex, a predator that luckily no human ever saw alive. Only fossils now remain of these mighty beasts.

History and Prehistory

TURNING THE PAGES OF TIME

History is the study of past human events. It deals only with the last 6,000 years or so—since writing was developed. Most people can expect to live for 70 or 80 years, longer than most animals. Looking at times past, we speak about "centuries" (100 years), or "ages," such as the Middle Ages (the period from A.D. 500 to 1500). A millennium (1,000 years) seems a very long time—with enough history to fill many books—yet history tells only the most recent part of the world's story.

Evidence from prehistory includes stone tools made by early humans. This is a modern replica of a Stone-Age ax made from flint with a wooden handle.

EXPLORING PREHISTORY

People began to use writing as a way to record events less than 6,000 years ago. To find out what happened before, we must turn to archeology.

Archeologists study the things people leave behind, such as stone buildings, iron tools, pieces of clay pottery, food remains, and the bones in their graves. From finds like these, experts can put together the story of how people became farmers and traders (about 11,000 years ago), how they built towns and began to create what we call civilization.

Even further back in time—before civilization —the clues are fewer and harder to find. But they do exist and tell us about people who lived as hunters and plant-gatherers in what we call the Stone Age. These people left no buildings or broken pots, but we can look at paintings they made in caves and even handle their stone tools.

Paintings made by Stone-Age artists, like these of wild animals on the walls of the Lascaux Caves in France, provide clues of prehistoric life. They were painted about 17,000 years ago.

THE FIRST HUMANS

Scientists have found a few clues—a bone, a pebble used as a tool, a footprint—left by very early humans who lived about two million years ago. Remains have also been found of even older humanlike creatures, such as *Australopithecus*, which lived in Africa more than four million years ago.

The human story began on the plains of Africa. Old though it is, this story starts only on the last page of the Earth-time book. To fill in the many blank pages before that, we need to hear the much older story, which is locked in the secrets of ancient rocks.

CHAPTERS IN THE EARTH-TIME STORY

Geologists divide Earth time into eras, each lasting millions of years. The eras are divided into periods, each with its own story. The most recent period is split into shorter time zones, called epochs. The oldest rocks containing fossils are known as Cambrian rocks, and rocks even older are called Precambrian. (Cambria is an old name for Wales, where very old rocks have been found.)

PRECAMBRIAN TIME *(see pages 10–13)*
From the beginnings of the Earth (4.6 billion years ago) until 570 million years ago.

3.3 billion years ago	Life begins in the oceans, with the first bacteria
3 billion years ago	First blue-green algae (releasing oxygen)
1.5 billion years ago	Eukaryotes (single-celled living things—the ancestors of all plants and animals)
1.1 billion years ago	Corals, jellyfish, and worms in the seas

PALEOZOIC ERA—"ANCIENT LIFE" *(see pages 14–19)*

CAMBRIAN PERIOD **570 million years ago**	Trilobites, mollusks, first jawless fish
ORDOVICIAN PERIOD **500 million years ago**	Trilobites, corals, mollusks; colonies of small animals called graptolites
SILURIAN PERIOD **435 million years ago**	First land plants; coral reefs in the seas
DEVONIAN PERIOD **410 million years ago**	Swampy forests on land; first insects and amphibians; bony fish and sharks
CARBONIFEROUS PERIOD **360 million years ago**	Many crustaceans, fish, and amphibians; trilobites dying out; tree ferns on land; first reptiles; giant insects
PERMIAN PERIOD **290 million years ago**	Cone-bearing trees—the first plants with seeds; many reptiles, fish, and insects

MESOZOIC ERA—"MIDDLE LIFE" *(see pages 20–24)*

TRIASSIC PERIOD **240 million years ago**	Crocodiles, turtles, first dinosaurs; first mammals
JURASSIC PERIOD **205 million years ago**	Cone-bearing trees still plentiful; large sea reptiles; giant dinosaurs; first birds
CRETACEOUS PERIOD **138 million years ago**	First flowering plants; many worms and insects; horned and armored dinosaurs; dinosaurs die out at end of Cretaceous

CENOZOIC ERA—"MODERN LIFE" *(see pages 25–29)*

TERTIARY PERIOD **65 million years ago**	
Paleocene Epoch	Flowering plants; reptiles and mammals thrive, so do insects and other invertebrates
Eocene Epoch	Birds, amphibians, small reptiles, and fish; first bats, camels, cats, horses, monkeys, rhinoceroses, and whales
Oligocene Epoch	First apes; camels, cats, dogs, elephants, horses, rhinoceroses, and rodents develop; some giant land mammals die out
Miocene Epoch	Apes spread; bats, monkeys, whales, first bears, and raccoons; trees and plants like those of today
Pliocene Epoch	Many animals are like modern kinds; first humanlike creatures
QUATERNARY PERIOD **2 million years ago to the present**	
Pleistocene Epoch	Mammoths and woolly rhinoceroses flourish; first human beings
Holocene Epoch	Humans hunt some animals to extinction, and tame others; humans change environments through farming, building, and other activities; environmental changes threaten habitats and wildlife

Rock Record

HOW IT ALL BEGAN

The story begins with a new world formed about 4.6 billion years ago. It was our Earth, one of nine planets that travel around a star we call the sun. This family of planets, the solar system, was probably formed from a huge whirling mass of gas and dust called a nebula, one of many nebulae in the universe. The universe was already old. Most scientists think it was created by a vast explosion of energy—the "Big Bang"—about 13 billion years ago.

According to a theory first put forward in 1905, tiny clumps of space matter were drawn together by gravity, to form miniplanets. Bigger lumps attracted smaller lumps that crashed into them, making the miniplanets even larger until they became the planets that today orbit the sun.

Whether the Earth began as a solid lump or (as another theory suggests) as a ball of gas, it eventually became a rocky mass surrounded by gases. Rocks are made from chemicals called minerals, in various combinations.

 A nebula is a star factory. This vast glowing cloud of dust and gas is the Orion Nebula, first seen in 1610 and more than a thousand light years away.

ROCKS AND PLATES

There are three kinds of rock. Sedimentary rocks are formed when other rocks, or even animal remains, get squashed together in layers. Igneous rocks are formed when melted rock, or magma, from the very hot interior of the Earth, cools and hardens. Both kinds can be chemically changed by great heat or pressure, becoming metamorphic rocks.

The Earth has three main layers: the crust (a thin layer of rock), the mantle (which contains some melted rocks), and the core (which has a liquid metal outer layer and a solid metal center). The Earth's surface is changing all the

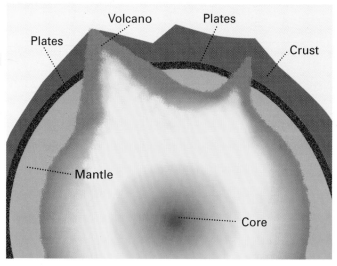

The Earth, basically a ball of rock, is very hot at its center, or core. The surface is constantly shifting, while volcanoes act as "escape valves," releasing enormous pressures deep inside the Earth.

time because it is made up of huge rocky plates that float on the hot molten rocks beneath. There are about 30 plates, like pieces of a giant jigsaw puzzle, which slide around at about four inches (10 cm) a year.

As the shifting plates bump against one another, they cause earthquakes and volcanoes, push up mountains, and make the continents "drift." Two colliding plates are pushing up the Himalayas, the world's highest mountains, which have been growing for the past 40 million years.

Another slow process wears away rocks. This is called erosion. Natural forces including wind, water, heat, ice, and chemical action smooth and grind down mountains, fill in valleys, and make rivers appear and disappear.

THE FOSSIL CLUES

Rocks give the first clues about how life began on Earth more than three billion years ago. Fossils of primitive bacteria are contained in rocks of this age.

A fossil is any part of an animal or plant preserved in rock. More than 2,000 years ago, Greek scientists were puzzled by fossils. In the 1020s, the great Arab scientist Avicenna decided that fossils must be failed attempts to create new animals and plants from stone. Other people thought that fossils were the remains of creatures drowned in the Great Flood described in the Bible.

By the 1800s, scientists realized that fossils told a different story. Dinosaurs had once walked the Earth, but there were no dinosaurs alive now. New kinds of animals had "evolved" to replace them. How had this come about?

A fossil ammonite. These shelled sea animals were common in the oceans of 200 million years ago, and many fossils have survived.

KEY DATES IN DISCOVERY

▶ **400s B.C.** Herodotus, a Greek historian, says fish fossils found in Egypt prove part of the land must once have been covered by sea

▶ **A.D. 79** Roman rock and fossil writer Pliny is killed while observing the erupting volcano Vesuvius

▶ **1669** Nicolaus Steno of Denmark discovers that some rocks form layers, and that the oldest layers are often the deepest

▶ **1780s** William Smith of Britain finds similar fossils in rock layers from different places; James Hutton, another British scientist, declares in 1785 that hot lava from volcanoes forms rocks, and that the world is constantly changing

▶ **1858** Charles Darwin and Alfred Russel Wallace put forward their theories of evolution by natural selection—the "survival of the fittest"

▶ **1912** Alfred Wegener of Germany suggests that "continental drift" explains why the continents are the shapes they are (see page 24)

Life Dawns

CREATION TIME

Most religions have a creation story, which tells how life on Earth began. The book of Genesis in the Bible describes how God made the Earth and all living things. Until the 1800s, few people doubted the truth of such stories. They believed all animals and plants had been created at the same time and that fossils were the remains of animals that had not been saved from the Flood by Noah and his Ark.

 The animals enter the Ark. Before 1850, few people looking at this painting of the biblical scene would have doubted the truth of the story.

LOOKING MORE CLOSELY

Scientists began to question these creation stories after the invention of the microscope in the 1600s. This revealed a new world. People had not guessed there were so many living things too small to be seen with the unaided eye. By the 1800s, scientists had begun to look more closely at fossils too and realized that living things change. So how had life begun, and how long ago?

This microscope was used by Robert Hooke (1635–1703), one of the first scientists to study microscopic life.

HOW LIFE BEGAN

Life on Earth began about 3.5 billion years ago, but exactly how is still a puzzle. The young Earth (about a billion years old) was turbulent. There was no water, but there were lots of chemicals, mixing together as the planet cooled. Energy from sunlight and lightning could have stirred up the chemical "soup" to make molecules of sugars and amino acids. These chemicals were able to join together, producing more complex molecules which could make copies of themselves.

Space, too, may have played a part. Comets and meteorites could have "seeded" life-giving chemicals into the planet's bubbling mixture.

Unlike the Earth, the moon (right) has no atmosphere or life. Passing comets (center) may have helped create life on Earth.

SEA LIFE AND AIR

From the oldest rocks, about 3.9 billion years old, comes evidence of water. Water was released from minerals, and also perhaps by comets (which are ice balls) passing close by or hitting the Earth. Water collected to form oceans on the Earth's surface. The oceans were a good breeding ground for some molecules. They multiplied and became primitive bacteria feeding on chemicals.

Over the next billion years, new living things appeared. Made of just one cell, they were the ancestors of today's plants and animals. They used energy from sunlight to make their food, giving off oxygen at the same time. This extra oxygen began to enrich the atmosphere, making air. Some oxygen became ozone, forming a layer that shielded the Earth's new life from the burning ultraviolet rays of the sun. Sheltered within the planet's blanket of air, and nursed in its warm, food-rich oceans, life began to multiply.

The First Animals

AN EMPTY WORLD

Imagine that we have traveled back in time more than a billion years. What do we see? The land is empty. There are no plants, just bare rock, mud, and sand. But peer into the sunlit oceans and we see creatures floating in the water or wriggling along the ocean floor. Some we recognize, but others look very strange to our eyes.

We have few clues to tell us what these early animals looked like. Their bodies were soft, so hardly any fossils survive. Since the appearance of hard-shelled animals, the fossil record has become easier to read.

 Jellyfish were abundant in prehistoric seas and have changed very little in millions of years. They were among the first success stories of life on Earth.

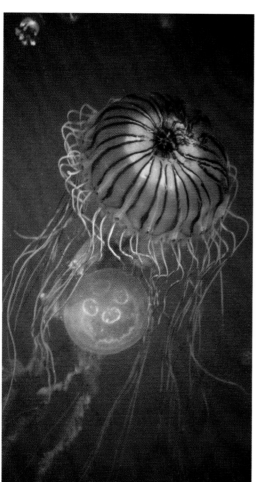

SHELLS AND SOFT BODIES

The first shelled animals were brachiopods, which looked like cockles and lived on the seabed. There were shelled mollusks, such as snails and starfish. Soft-bodied worms and jellyfish were plentiful, as were sea lilies—flowerlike animals that resembled starfish on the end of long stalks.

Along the sea floor crawled animals like giant woodlice. These were trilobites, with jointed legs like crabs. Trilobites were very common. Fierce sea scorpions, as long as a person, scuttled over the sand and rocks, chasing their next meal.

A fossil trilobite, found in the Burgess Shale, a rich fossil-bearing region of black slate-type rock in western Canada. This animal once wandered about the sea floor.

THE FIRST FISH

None of these early creatures had a skeleton made of bone. Their bodies were protected by hard shells, like beetles and crabs today.

The first animals with bony skeletons (known as vertebrates) swam in the oceans 500 million years ago. These animals were the first fish, but were unlike the fish you see in an aquarium today. For example, they had no jaws, but instead had sucking mouths. They were covered in bony armor and moved slowly along the sea floor, guzzling scraps of food. Some grew large and were fierce hunters, grabbing prey in jaws armed with bony plates instead of teeth.

About 420 million years ago, the first fish with jaws appeared. Among them were sharks and rays, and fish that looked like sturgeon. The new fish ruled the seas in what is often called "the Age of Fishes."

By 200 million years ago, fish looked much as they do today. They had scales instead of heavy armor, and spiny fins for swimming and steering. There were lots of ammonites, too. These large mollusks had coiled, chambered shells rather like the nautilus of today.

The coelacanth is a primitive fish thought extinct until a live specimen was caught off South Africa in 1938. We now know coelacanths live on both sides of the Indian Ocean.

KEY DATES: LIFE IN THE OCEANS

▶ **570 million years ago**	First trilobites
▶ **500 million years ago**	Colonies of small animals called graptolites; mollusks and corals; plentiful algae; first jawless fish
▶ **435 million years ago**	Corals form reefs in oceans
▶ **420 million years ago**	First fish with jaws and bony skeletons
▶ **410 million years ago**	Many kinds of jawed fish; sharks develop
▶ **345 million years ago**	Sea lilies form vast colonies
▶ **245 million years ago**	Ammonites plentiful
▶ **200 million years ago**	Modern-looking fish swim in the oceans

Amphibians Arrive

THE MOVE TO DRY LAND

The next important date on the Earth-time calendar came about 430 million years ago. Plants—until now water-living—spread to dry land. The first land plants were probably small, greenish-brown, rubbery stalks, clinging to wave-splashed rocks. They spread up beaches beyond the high-tide mark and along the banks of river estuaries.

Until plants began to grow on the shore, the animals could not follow, because there would have been nothing to eat on land. Soon after the plants came the first land animals, which were small creatures that could breathe air through their hard skins and crawl about on stiff legs. These were the ancestors of insects and spiders. Could larger animals follow?

LAND EXPLORERS

As plants spread, they made life on land easier for animals because they provided shelter. More importantly, they gave off oxygen that land animals could breathe, once they developed air-filled lungs to replace the fishy gills that worked only in water.

Some sea creatures may have crawled from the water to escape pursuing enemies. Others crept ashore to feed on the green plants and small insects plentiful at the water's edge. For these explorers, life on land was better.

Fossil ferns from coal seams in England. During the Carboniferous period 360 million years ago, swampy forests of ferns and other plants decayed and began to form the coal we burn today.

FISH WITH LEGS . . .

Crawling fish were the first backboned animals to explore land. Such fish lived close to shore and in tidal pools. Like the modern African lungfish, they could spend some time out of water, perhaps burying themselves in mud burrows.

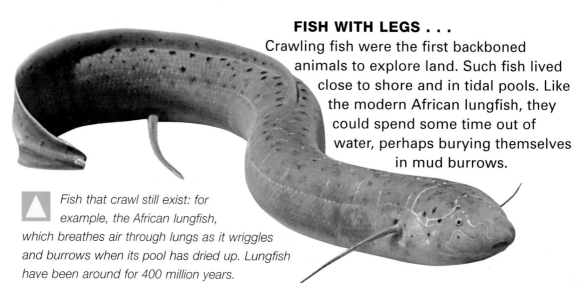

Fish that crawl still exist: for example, the African lungfish, which breathes air through lungs as it wriggles and burrows when its pool has dried up. Lungfish have been around for 400 million years.

The first land fish used their stiff fins for crawling as well as for swimming. They came from the water to feed and to escape predators. At first they made only short visits to land, but after a while they developed air sacs in the throat and could breathe air more easily. This allowed them to stay longer on land. They returned to water to lay their eggs.

. . . BECOME FROGS

Time, through the action of evolution, was changing these animals. They had primitive lungs, not gills. Their fins had become stubby legs. Thick skins kept them from drying out in the warm sun. They were creatures of land and water—the first amphibians.

An Australian tree frog. Frogs evolved from amphibians with tails, rather like today's newts and salamanders. As amphibians go, frogs are the last word in development.

ANCIENT AMPHIBIANS

Most of the earliest amphibians, such as Ichthyostega, looked like large newts with fishy tails. Their feeble jaws could just manage to grab a passing insect, although Eogyrinus—a kind of large salamander—probably had a nasty bite. So had Eryops, a crocodile-like creature more than three feet (1 m) long. It gobbled up other amphibians and fish. There were some very large amphibians, including Mastodonsaurus, which was about 15 feet (4.6 m) long. The earliest known frog—Triadobatrachus—had a short tail. Modern frogs and toads have lost their tails, except when they are tadpoles.

KEY DATES: AMPHIBIANS ADVANCE

▶ **370 million years ago**	First known fossils of amphibians
▶ **300 million years ago**	Lots of amphibians
▶ **270 million years ago**	Large amphibians like Eryops flourish
▶ **220 million years ago**	Earliest frog fossils

Land Life

GIANT CRAWLERS AND FLIERS

For a long time, the amphibians and invertebrates (worms, spiders, and insects) had the land to themselves. Some became giants. There were amphibians as long as cars, and insects as big as rabbits. Cockroaches four inches (10 cm) long scuttled over the ground, while dragonflies the size of pigeons zoomed through the forest branches. It had taken 300 million years to travel from the trilobites and shellfish of the Cambrian oceans to slow-crawling and not very intelligent amphibians.

The cockroach is one of nature's great survivors. A time traveler into the prehistoric past would find plenty of these persistent pests scuttling around.

A fossilized dinosaur egg, laid some 75 million years ago but never hatched. Reptiles had to adapt to breed on land rather than laying eggs in water, like amphibians.

REPTILES TAKE OVER

From a time machine set to explore the world of 330 million years ago, we would see different creatures. Amphibians were no longer dominant. The new top dogs looked like amphibians, but they needed water only for drinking and they laid their eggs on land. They were the first reptiles.

Reptiles were cold-blooded, like amphibians, but with a useful advantage: their eggs had tough shells, which kept the baby inside from drying out even when buried in hot sand. Reptiles also had longer legs, so they ran and climbed better than amphibians. They moved away from lakes and rivers to take over new habitats, such as deserts and mountains.

REPTILES, SMALL AND BIG

Evolution produced an interesting mixture. Some early reptiles—such as Hylonomus—were like small lizards, fast-running and with sharp teeth for catching insects. The bigger, sail-finned Dimetrodon was a meat eater. Plant-eating Moschops was a lumbering creature as big as a cow, and may have lived in a herd, as cows do today. Most reptiles ran or crawled on all fours, but some took to walking on two legs.

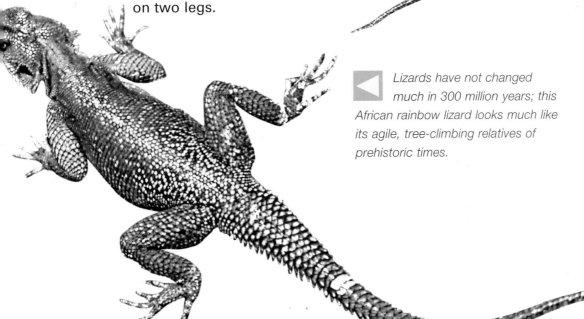

Lizards have not changed much in 300 million years; this African rainbow lizard looks much like its agile, tree-climbing relatives of prehistoric times.

KEY DATES: REPTILE REVOLUTION

▶ **330 million years ago** First reptiles develop from lizard-like amphibians
▶ **290 million years ago** The Earth becomes warmer and drier; reptiles spread
▶ **240 million years ago** "Age of Reptiles" begins
▶ **205 million years ago** Large amphibians die out, but small ones such as frogs and salamanders survive; turtles, crocodiles, pterosaurs, and dinosaurs evolve

Dead Bones Tell Tales

Reptiles were superior for 250 million years. They were not the brightest of creatures. Many primitive reptiles had tiny brains, so they were probably slow-witted, as well as slow-moving. But others, especially the more agile hunters, must have been more intelligent.

Cynognathus was a dog-sized hunter that was more like a mammal than a reptile. It had dog-like teeth, and probably a hairy body. Scientists have guessed this because tiny root holes for whiskers can be seen in fossil Cynognathus skulls.

A dragon at the beach. The world's largest lizard, the Komodo dragon from Indonesia, is a midget compared with the giant extinct dinosaurs—just as well, for even this midget is big enough to eat a pig.

NEW REPTILES MOVE IN

Near the end of the Permian period, about 240 million years ago, there was a mass extinction of animal life. Many species, especially sea animals, died out. In their place, new animals emerged in the seas and on land. These newcomers included large sea reptiles, turtles, crocodiles, and the first true mammals, which looked like shrews, and scuttled out of the way of larger and more formidable animals, such as dinosaurs. Dinosaurs, in many forms, were to rule the Earth for 140 million years (see pages 20–23).

Turtles and tortoises survived the mass extinctions and environmental changes that wiped out the dinosaurs and marine reptiles such as Ichthyosaurs.

BONES BAFFLE EXPERTS

In 1677, part of a dinosaur bone was found near Oxford in England. Scientists declared it was either an elephant's bone, or had belonged to a human giant! During the 1700s, many bones of prehistoric animals were found in Europe and America, but scientists disagreed about how they came to be inside rocks. Even people who considered themselves to be experts were unable to tell the difference between animal bones and bones from human beings. The bones of a prehistoric salamander found in Germany in 1725 were said to be those of a person, drowned in Noah's Flood.

THE FOSSIL PUZZLE

It had taken millions of years for dinosaurs to evolve into dinosaurs. We can piece together the story from scattered clues in the form of fossils.

Fossils puzzled people. As long ago as the sixth century B.C., two Greek scientists, named Thales and Anaximander, looked at fossils of fish. They decided that these "stone bones" belonged to animals that lived long ago.

People went on finding fossils. Farmers dug them up while plowing fields. Stone cutters found them as they hacked rocks from quarries. Some were called "dragon's teeth." They were probably fossil belemnites, prehistoric squid, but because of the many stories of magicians turning people to stone, those who found them wondered if they were parts of long-dead giants!

This dinosaur thigh bone is being recovered from 200-million-year-old sandstone rock. No wonder dinosaur bones were once mistaken for giant's limbs!

Dinosaur Times

BONES, JUST BONES

No human has ever seen a live dinosaur; perhaps that is why people find them so fascinating. There were many types of these ancient reptiles, some unlike any animals we can see today: a number of them were

giants, while others were no bigger than chickens. Dinosaurs vanished from the Earth 65 million years ago; we know about them only from bones, footprints, and a few fossilized eggs. The existence of some dinosaurs is evident from just a few teeth or fragments of bone. Piecing together the remains, an expert can make a lifelike model of the dinosaur. Some features, such as skin color, leave no clues in the rocks and so must be guessed at.

The claw print of a dinosaur, left in soft mud which then hardened into rock. Such footprints give clues as to how dinosaurs stood, walked, and ran.

Discovering dinosaurs changed the way scientists looked at time, and the way ordinary people imagined it too. Dinosaur bones caused great excitement when they were first collected and studied seriously in the early 1800s. What could these creatures be? When had they lived? What did they look like?

The largest complete Tyrannosaurus rex skeleton yet found is on show in Chicago's Field Museum. The powerful back legs provided the speed, the long tail the balance, and the massive skull the killing power.

WHAT'S AN ERA?

To visit the world of the dinosaurs, we must go back 240 million years, to the Mesozoic era. Eras are huge chunks of time. The Mesozoic era lasted about 180 million years. Its name means "middle life" because it came after the Paleozoic ("ancient life") and before the Cenozoic ("modern life") eras (see page 7).

Like other eras, Mesozoic is split into periods: the Triassic, the Jurassic, and the Cretaceous. The Mesozoic era was a lively time for nature: birds, large sea reptiles, and flowering plants were all developing at this time.

THE DINOSAUR HUNTERS

One of the first fossil collectors was Mary Anning (1799–1847), who as a child found the first complete fossil skeleton of the sea reptile Ichthyosaurus in the chalk cliffs of Lyme Regis in Dorset, England. A French scientist, Baron Georges Cuvier, became the first expert at putting together complete fossil skeletons. Museums became so anxious to have "prehistoric monsters" that fossil hunters led large expeditions to dig up bones. American fossil hunters Othniel C. Marsh and Edward Cope found new species of dinosaurs and prehistoric mammals in the American West. They became fierce rivals, their teams practically fighting to reach a site first.

▲ *An artist's impression of an Allosaurus, a Jurassic dinosaur similar to the later Tyrannosaurus, but slightly smaller. This monster roamed North America.*

DINOSAUR TAKEOVER

At the start of the Mesozoic era, the primitive reptiles were still flourishing, but within 20 million years, they had died out. Maybe the climate changed or competition from the newcomers was too strong. Whatever the reason, the way was clear for a Triassic takeover. In came the dinosaurs, such as Eoraptor and Procompsognathus—a hunter that ran swiftly on its back legs.

Gentle Giants?

The dinosaurs lived through three periods of Earth time: the Triassic (35 million years), Jurassic (67 million years), and the Cretaceous (73 million years). Some of them were giant-sized. Largest of all were the massive, long-necked Sauropods, such as Brachiosaurus, Apatosaurus, and Ultrasaurus. These peaceful plant eaters were easily the largest land animals that have ever lived. Weighing 70 tons (64 t) or more, they were tall enough to see over the roof of a house—had there been any houses. The hunters that preyed on these giants were just as impressive, though not so large.

No teeth ever looked more deadly than the 16-inch (40 cm) razors of the Tyrannosaurus rex. Its mouth was big enough to swallow a person whole!

CRETACEOUS KILLERS

Many of the best-known dinosaurs, such as Tyrannosaurus rex and Velociraptor, lived in the Cretaceous period. This was late in dinosaur time: they were "modern" dinosaurs.

Any time traveler would be wise to keep clear of Tyrannosaurus (meaning "tyrant lizard"). This was the most fearsome predator of the dinosaur age: a 45-foot-long (14 m) monster, with jaws that could savage even armored Triceratops ("three-horned face").

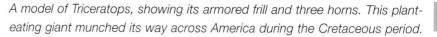
A model of Triceratops, showing its armored frill and three horns. This plant-eating giant munched its way across America during the Cretaceous period.

Velociraptor ("speedy thief"), no bigger than a large dog, crippled victims with slashes of its hind claws. Just as ferocious was the man-sized Deinonychus ("terrible claw"), which probably hunted in packs, like wolves today.

SWIMMERS AND FLIERS

At the time when dinosaurs roamed the land, large sea reptiles swam in the seas, hunting fish. Dolphin-like Ichthyosaurs and long-necked Plesiosaurs both flourished during the Jurassic period.

The air above them was full of strange flying shapes: reptiles gliding on bat-like wings. The biggest of the Pterosaurs, Quetzalcoatlus, was the size of a small airplane, with wings 33 feet (10 m) across.

Quetzalcoatlus was the biggest of the Pterosaurs, a group of flying reptiles that flourished in the Jurassic and Cretaceous periods. The biggest flying animal ever, it probably glided on rising currents of air.

KEY DATES: DISCOVERING DINOSAURS

▶ **1800**	The first dinosaur footprints are found in North America
▶ **1820s**	Fossil bones of Iguanodon (a dinosaur) and Ichthyosaurus (a sea reptile) found in England; the age of fossil hunting begins
▶ **1822**	Mrs. Mary Mantell finds the tooth of an ancient animal in Sussex, England; her husband, Dr. Gideon Mantell, names the animal Iguanodon
▶ **1824**	Professor William Buckland describes some bones he is given as belonging to a giant reptile; more reptile bones are found
▶ **1841**	Richard Owen suggests the name dinosaur ("terrible lizard") for the long-dead reptiles
▶ **1853**	Life-size dinosaur models go on show in London; the dinosaur craze has begun
▶ **1860**	A fossil feather from Archaeopteryx, one of the earliest birds, is found in Germany; then a whole skeleton is found
▶ **1880s**	American dinosaur hunters Edward Cope and Othniel C. Marsh are bitter rivals, traveling the West in search of fossils

Mammals Move In

WAITING FOR THEIR CHANCE

Earth time is not cut into clear "chapters" as it may appear in timelines in books. The dinosaurs did not disappear one day and mammals take over the next. Mammals had lived on Earth for almost as long as the dinosaurs. They were small animals, mostly active at night when the dinosaurs were sleeping. Being warm-blooded and furry-coated, the little mammals stayed lively even after the sun set. Birds evolved from dinosaurs, and they too were warm-blooded, but with feathers instead of fur.

EARTH'S CHANGING PATTERNS

While dinosaurs munched leaves (or each other), and the little mammals scurried about in the moonlight, time was not standing still. The planet changed constantly.

The continents as we know them (Africa, North America, South America, Asia, Europe, Antarctica, Australasia) did not exist in dinosaur times. All land was joined together, first as a giant supercontinent, then as two megacontinents. By about 100 million years ago, the continents were looking more like they do today. Animals that had once been able to mingle freely were now separated by oceans and confined to their continents.

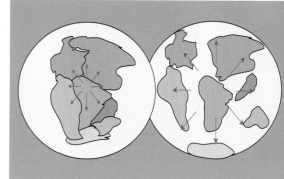

Changing continents. The prehistoric supercontinent, named Pangaea, broke in two, and the land masses then slowly drifted apart to take the shapes we know today. This movement is still going on.

THE END OF THE DINOSAURS

Something happened 65 million years ago that is thought to have wiped out the dinosaurs: possibly an asteroid hurtling through space smashed into the Earth. The huge impact started fires that sent clouds of dust and smoke into the air. The clouds hid the sun, and a terrible gloomy winter lasted for years. Many plants died. With nothing to eat, the giant plant-eating dinosaurs starved to death. The meat eaters soon followed.

A meteor crater in the Arizona Desert. A much bigger space impact, from a crashing asteroid, may have destroyed the dinosaurs.

CATASTROPHE OR CLIMATE CHANGE?

Changes in climate and plants could also have killed off the dinosaurs. Unable to digest new plants, or feed their young, dinosaurs quite quickly died out. There have been other mass extinctions of animals. But why the so successful dinosaurs (and not other reptiles such as crocodiles) died out is still puzzling.

MULTIPLYING MAMMALS

The mammals multiplied to replace the dinosaurs. Mammals as different as mice and elephants existed only 10 million years after the dinosaurs had disappeared. There were water mammals (whales) and flying mammals (bats).

Time switched evolution into a higher gear, speeding up change. Mammals appeared and, if they were not adaptable enough, quickly disappeared: six-tusked elephants, saber-toothed cats, giant flightless birds, giant kangaroos. The most advanced animals were the placental mammals, which gave birth to developed, live young.

TWO LEGS, TWO HANDS

Lemurs were the first primates, or humanlike mammals, and they appeared about the time the dinosaurs died out. The "tree of life" had sprouted a new branch. After many millions of years came apes, and later ape-like creatures which walked on two legs, using their hands to grip sticks and stones. A new age in Earth's story was beginning.

Lemurs were the first primates but now live only on the island of Madagascar. More advanced primates (monkeys and apes) have taken their place everywhere else.

KEY DATES: BIRDS AND MAMMALS

▶ 138 million years ago	The first birds appeared; most modern insect and fish groups are thriving
▶ 65 million years ago	Dinosaurs die out; "Age of Mammals" begins
▶ 55 million years ago	First camels and horses
▶ 38 million years ago	Grasslands spread, as the climate becomes cooler and drier; grass-eating animals such as deer evolve
▶ 30 million years ago	The first apes live in forests
▶ 24 million years ago	Mammals are most plentiful and varied
▶ 5 million years ago	First humanlike creatures, walking on two legs
▶ 2 million years ago	Ice Age begins; woolly mammoths and other creatures adapt to cold conditions

The Evolution Clock

WHAT IS EVOLUTION?

Evolution is a slow process of change that affects all living things. Evolution works by adapting things that already exist. A bat's wing and a dolphin's flipper look very different. But their "arm" and "finger" bone structure is so similar that it is likely these animals share a common ancestor. Most scientists believe that evolution explains how living things come to look as they do today.

In the 1850s, the scientists Charles Darwin and Alfred Russel Wallace suggested that animals of the same species can pass on variations to their young. A useful variation, such as a bigger beak in a seed-cracking bird, gives "big-beaked" offspring a better chance of survival. In time, by "natural selection," big beaks become common. A new species has evolved.

 Charles Darwin, who sailed around the world (1831–1836) collecting evidence for his new theory.

TIME "MODELS" SPECIES

You might look like your parents or your grandparents. All animals and plants inherit size, coloring, speed, food likes, and other characteristics from their parents. They are copies of the same model.

However, living things also inherit the ability to change. They change over time to make better use of the changing conditions in their environment (the place and climate in which they live). Let's use the horse as an example: its ancestors lived in forest undergrowth and had short legs with four toes; modern horses live on grasslands, where to escape enemies, they run fast on long legs and only one toe (a hoof).

Studies of chimpanzees and other apes show how their behavior mirrors our own in many ways. In 1871 Darwin suggested that apes and humans shared a common ancestor. This idea shocked many people.

EVOLUTION PICKS THE STRONGEST

Those animals and plants best suited to their environment are the strongest. They produce the most offspring, so their good qualities are passed on.

If the environment changes, plants and animals that can also change—by finding new foods or places to live—are most likely to survive. Those that cannot change die out.

TIME AND THE LANDSCAPE

The environment changes constantly over time. The wet, swampy fern forests of the prehistoric amphibians were replaced by cooler, drier tree forests. Forests gave way to grasslands. Some grasslands became deserts. In some places, sea covered the land. In others, new land rose from beneath the waves.

If we traveled back just 20,000 years in time, we would not recognize our world. There would be no cities, no roads, little sign of human life. A few blackened patches of burned-out forest might be all there was to show that humans had begun to make their own changes to the world.

The natural rain forest in Puerto Rico. Such landscapes change naturally, but human activity (road building, tree cutting, farming) can lead to sudden environmental disaster.

KEY DATES

▶ **1809** Professor Jean Lamarck of France puts forward a theory of evolution

▶ **1822** Georges Cuvier of France studies fossils and suggests that animals can change

▶ **1830s** The Swiss-American scientist Louis Agassiz explains how glaciers and ice sheets change the Earth's surface

▶ **1830** British scholar Sir Charles Lyell writes the first modern textbook of geology—the study of the Earth's rocks

▶ **1846** Robert Mallet of Ireland begins the scientific study of earthquakes

▶ **1859** British naturalist Charles Darwin publishes his study of evolution and "natural selection," *The Origin of Species*; his theory is backed up by the work of Alfred Russel Wallace

Where Will Time Take Us?

TIME FOR PEOPLE

The first people were few in number, wandering in small bands over the grassy plains of Africa. But their intelligence, and their ability to make stone tools and work together, made them stronger than they looked: they could hunt animals bigger than themselves.

THE FARMING REVOLUTION

The first people did not look quite like us. It took two million years for "modern" people to develop. During that time, humans lived alongside other animals, often competing with them for food and territory. Humans were evolving with time, making them more intelligent, more organized, and more powerful.

Forty thousand years ago, people were spreading out of Africa across Europe and Asia. By 11,000 years ago, some people had stopped wandering as hunters and settled down as farmers. They tamed and bred sheep, cattle, and pigs. They rode horses, donkeys, and camels. Farmers began to change the landscape by clearing land for fields, planting crops, such as wheat, where there had been grass, and by cutting down trees for firewood and to make buildings. They built towns, made roads, and dug canals to water their fields.

The Industrial Revolution introduced factories and pollution, which created smoky landscapes as shown in this 19th-century painting of the Black Country, the West Midlands, England.

QUICK TIME, WASTE TIME

For thousands of years, these changes affected only small areas of the planet. Much of the Earth remained untouched by people. A thousand years ago, there were no more than 400 million people on Earth. That is less than half the population of India today. There are now more than six billion people on Earth.

The 1700s saw a period of rapid change which we now call the Industrial Revolution. With more food and better medicines, people lived longer, and more children survived childhood and grew up. The population began to grow much faster. To feed new industries, people used the Earth as never before—mining for coal, digging for minerals, drilling for oil.

THROUGH THE "FUTURESCOPE"

Recent time has not been kind to nature. If we could peer through a "futurescope," what changes might we see?

We might see a world with few wild places and few wild animals. This world would be one of traffic-clogged cities, polluted rivers, and garbage-littered

Raw materials, such as metals, and fuels, such as coal and oil, will not last forever. We must turn to alternative energy sources such as wind power, which is harvested by wind turbines.

oceans—a world running out of time as greedy humans use up its resources. Climate changes may bring warmer winters to some people, but worsening floods and storms to others.

On the other hand, we could see a much cleaner world, which is sensibly managed so that humans share the Earth with the rest of nature. In this future world, people keep air and water clean, protect and plant forests, leave spaces for wildlife, and use energy from reusable sources, such as the wind, sun, and ocean tides.

It is up to us which world we will see. The world changed greatly in the 20th century. How will it go on changing in the 21st? It's hard to imagine. Even in a world dominated by one species—humans—evolution will continue. Looking far ahead by thousands or millions of years, the Earth will be home only to those things which have been able to adapt and survive. How people will change, only time can tell. People of the future may even leave the Earth for outer space, to settle on other worlds and begin a new era of human time.

Busy Tokyo, Japan, is typical of a modern, overcrowded city. The bigger cities grow, the more countryside they swallow.

KEY DATES: THE RISE OF HUMANS

▶ **4 million years ago**   Ape-like *Australopithecus* lives in Africa

▶ **2 million years ago**   *Homo habilis* ("skillful human"), the first true human makes stone tools

▶ **1.75 million years ago**   *Homo erectus* ("upright human"), better at making stone tools, probably moves from Africa to other regions

▶ **300,000 years ago**   *Homo sapiens* ("wise human"), with large brains develops new skills such as building shelters

▶ **30,000 years ago**   Modern-looking people hunt, make clothes and tents, use fire, paint on cave walls

▶ **11,000 years ago**   People become farmers, begin building towns; human population numbers between 5 and 10 million

▶ **1,000 years ago**   About 400 million people live on the Earth

▶ **21st century**   Human population passes six billion, spreads across most of the planet

Glossary

Amphibian
An animal that can live on land and in water.

Bacteria
Tiny creatures made up of single cells.

Carboniferous
The "age of coal swamps," 360 million years ago; plants from this period became buried and changed over millions of years into coal.

Cell
The basic unit which makes up all living things.

Cenozoic
The "age of mammals," from 65 million years ago to the present.

Chemicals
Natural substances that can combine with each other; gases, salts, and acids are all chemicals.

Climate
The average weather of a region.

Continent
A large land mass, such as Africa or North America.

Cretaceous
"Chalk age," the period from 138 to 65 million years ago.

Epoch
A period of time shorter than an era. We are living in the Holocene epoch, which began 10,000 years ago.

Era
A long period of Earth time, lasting millions of years.

Evolution
Means "unfolding;" the development of living things over time, as a result of changes in their living conditions.

Extinction
The dying out of a group or species of animals or plants.

Fossil
The remains of a long-dead animal or plant, preserved in rock.

Geology
The study of the Earth through its rocks.

Habitat
The place where an animal or plant naturally grows.

Ice Age
A time, lasting thousands of years, when ice sheets covered much of the Earth's surface.

Jurassic
Named after the Jura Mountains in Europe, the period from 205 to 138 million years ago.

Mammal
A warm-blooded, hairy animal; it gives birth to young that feed on milk from their mother's body.

Marsupial
A primitive mammal; it gives birth to undeveloped young carried in a pouch on their mother's body.

Mesozoic
The "age of the dinosaurs," from 240 to 65 million years ago.

Mineral
Most common solid material on Earth. There are about 3,000 kinds but only about 100 common ones.

Molecule
The smallest amount of a chemical substance that can exist on its own.

Oxygen
A gas that animals breathe, present in the atmosphere; it is produced by plants during photosynthesis (using sunlight and water to make food).

Period
A division of Earth time, shorter than an era.

Permian
Named from the Perm Mountains of Russia; a period of time that lasted from 290 million to 240 million years ago.

Pollution
The fouling of air, water, or land by refuse, smoke, chemicals, and other products of modern life.

Reptile
A four-legged animal with scales that lays tough-shelled eggs on land; living reptiles include snakes, lizards, and turtles.

Resources
Anything from the Earth used by people, such as soil, water, minerals, and plants.

Rock
Combinations of minerals, forming the solid part of the Earth: there are three kinds of rock, called igneous, sedimentary, and metamorphic.

Skeleton
The bony framework that supports the body of mammals, birds, reptiles, amphibians, and most fish.

Species
A group of related animals that look alike and can breed with one another.

Triassic
"Three-fold," named after a German rock-system; the period from 240 to 205 million years ago.

Vertebra
A small bone in the spine or backbone; backboned animals are called vertebrates.

Index